First Picture Encyclopedia of Dinosaurs

by Belinda Gallagher
illustrated by Lucy Semple

ARCTURUS

This edition published in 2025 by Arcturus Publishing Limited
26/27 Bickels Yard, 151–153 Bermondsey Street,
London SE1 3HA

Copyright © Arcturus Holdings Limited

All rights reserved. No part of this publication may be reproduced, stored in a retrieval system, or transmitted, in any form or by any means, electronic, mechanical, photocopying, recording, or otherwise, without prior written permission in accordance with the provisions of the Copyright Act 1956 (as amended). Any person or persons who do any unauthorized act in relation to this publication may be liable to criminal prosecution and civil claims for damages.

Author: Belinda Gallagher
Illustrator: Lucy Semple
Consultant: Dougal Dixon
Editor: Violet Peto
Designer: Simon Oliver
Managing Editor: Joe Harris
Design Manager: Rosie Bellwood-Moyler

ISBN: 978-1-3988-3342-5
CH011703NT
Supplier 29, Date 1024, PI 00008266

Printed in China

Contents

When Were Dinosaurs Alive?	4
The Prehistoric World	6
Life on Land	8
The Dinosaurs Arrive	10
Ocean Dwellers	12
Fierce Hunters	14
Fliers and Gliders	16
Dinosaurs Rule the Land	18
How It Ended	20
Rise of the Mammals	22
How We Know	24
Peaceful Plant-Eaters	26
Altirhinus	28
Stegosaurus	30
Ankylosaurus	32
Psittacosaurus	34
Kosmoceratops	36
Torosaurus	38
Pachycephalosaurus	40
Ingentia	42
Diplodocus	44
Sauroposeidon	46
Argentinosaurus	48
Feathers, Beaks, and Claws	50
Citipati	52
Oviraptor	54
Deinocheirus	56
Compsognathus	58
Archaeopteryx	60
Deinonychus	62
Velociraptor	64
Mighty Meat-Eaters	66
Ceratosaurus	68
Allosaurus	70
Giganotosaurus	72
Spinosaurus	74
Tyrannosaurus Rex	76
In Air and Water	78
Quetzalcoatlus	80
Mosasaurus	82
Psephoderma	84
Archelon	86
Elasmosaurus	88
Sarcosuchus	90
Kronosaurus	92
Glossary	94
Pronunciation Guide	95
Index	96

When Were Dinosaurs Alive?

Earth has existed for a long, long time. For much of this time, there was no life—no plants or animals. But as conditions on Earth changed, simple life began to appear. Slowly, plants and animals spread around the world. The long stretches of time over which this happened are called eras. The eras then break into smaller chunks called periods.

THE PALEOZOIC ERA
(538–252 million years ago)

THE MESOZOIC ERA
(252–66 mya)

THE TRIASSIC PERIOD

THE JURASSIC PERIOD

Prehistoric Times

Prehistoric animals slowly changed. They developed new ways to find food and adapted to live in different places. They spread from water to land. It took many millions of years of gradual change before dinosaurs emerged. They lived during an era called the Mesozoic. The first dinosaurs appeared in the Triassic Period.

THE CRETACEOUS PERIOD

THE CENOZOIC ERA
(66 mya–Present)

The Prehistoric World

Long before the dinosaurs, the first animals lived in the ocean. As Earth slowly changed over millions of years, life changed, too. Animals that lived in water slowly adapted to find food and new places to live. Their bodies began to change to allow them to move and breathe on land.

Water and Land

Diplocaulus lived on land and in water like frogs, toads, and newts do today. It had an odd-shaped head that may have helped it to swim smoothly. The strange-shaped head may also have stopped other animals from eating *Diplocaulus*.

Acanthostega

Lungs and Gills

Acanthostega had gills and lungs so it could breathe in water and on land. It had a long body like a newt and sharp teeth for catching food.

Diplocaulus

306 to 255 mya

SAY IT: "dip-low-call-us"

HOME: Rivers, lakes, and swamps of North America

FOOD: Fish, insects, and other small water animals

BONUS FACT: I laid my eggs in water just like a frog does today.

Walking on Fins

The ancestors of *Diplocaulus* and *Acanthostega* were fish. *Tiktaalik* had long, strong fins that helped it wade nearer to land to catch insects and small animals to eat.

Diplocaulus

Tiktaalik

Life on Land

Animals began to live on land, and they continued to change. As some kinds died out, new ones adapted to take their place. By the end of the Paleozoic era, some animals even looked like dinosaurs. They had four legs, lizard-like bodies, and sharp teeth.

Before the Dinosaurs

Dimetrodon is sometimes mistaken for a dinosaur—but it is more closely related to humans! It lived millions of years before the dinosaurs. The spiny sail on *Dimetrodon's* back made it look big and fierce. This hunter had strong jaws and rows of long, stabbing teeth. It probably ate amphibians such as *Diplocaulus*.

Dimetrodon

Sprawling Legs

Varanops was a small hunter with sharp teeth. Its legs stuck out to the sides as it walked, and it had a long body like a lizard. It was an early type of reptile.

Varanops

Eating Plants

Moschops looked fierce, but it probably spent most of its time eating tough plants. It may have lived in small groups.

Moschops

Dimetrodon

295 to 272 mya

SAY IT: "die-MET-roh-don"

HOME: Around swamps and wetlands in Europe and North America

FOOD: Small reptiles, fish, and amphibians

BONUS FACT: My spiny sail may have helped me find a mate.

The Dinosaurs Arrive

The first dinosaurs were small and fast. Their legs were long and didn't sprawl out to the sides. This meant that they could chase after small lizards and mammals. Over millions of years, dinosaurs spread across the planet. New kinds appeared—from tiny meat-eaters to giant plant-eaters. All dinosaurs were reptiles.

Millions of Years Ago

The time when dinosaurs were alive is called the Age of Dinosaurs. It is divided into three periods, which each lasted for millions of years..

65 mya
Cretaceous
146 mya
Jurassic
210 mya
Triassic
245 mya

Moving on Two Legs

Herrerasaurus was one of the first dinosaurs. It could run quickly on two legs to hunt lizards and smaller dinosaurs.

Herrerasaurus

Small and Speedy

Another early dinosaur was *Coelophysis*. It moved quickly on two long legs with three-toed feet. Its snappy jaws were filled with sharp teeth. This small dinosaur had forward-facing eyes, so it could easily spot a possible meal. Sharp-clawed hands helped it hold down prey to eat.

Coelophysis
216 to 196 mya

SAY IT: "see-low-FIE-sis"

HOME: Dry plains of North America

FOOD: Lizards, insects, and small mammals

BONUS FACT: My leg bones were hollow, which allowed me to run fast.

Ocean Dwellers

Life in the oceans was very different during the Age of Dinosaurs. Ichthyosaurs were dolphin-shaped reptiles that swam beneath the waves. They had long, narrow jaws and powerful bodies for swimming fast. Like all reptiles, they had lungs, so they had to come to the surface to breathe air.

Fast Swimmer

One of the fastest ichthyosaurs was *Stenopterygius*. It had a powerful tail to push through the water as it hunted fish and squid to eat. This speedy swimmer had good eyesight to help it spot prey in deeper water, where there was less light. Like all ichthyosaurs, *Stenopterygius* gave birth to live babies and didn't lay eggs.

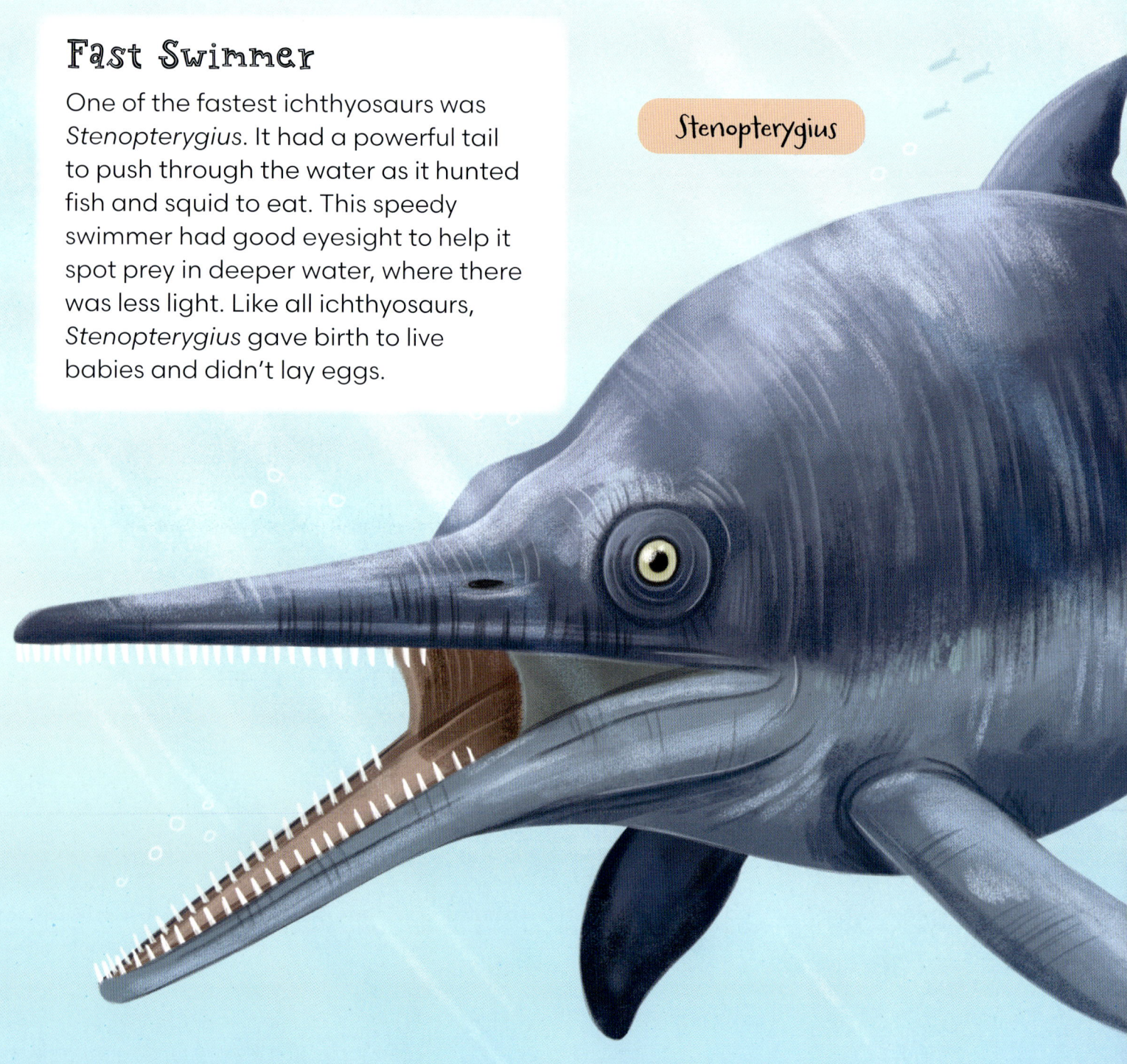

Stenopterygius

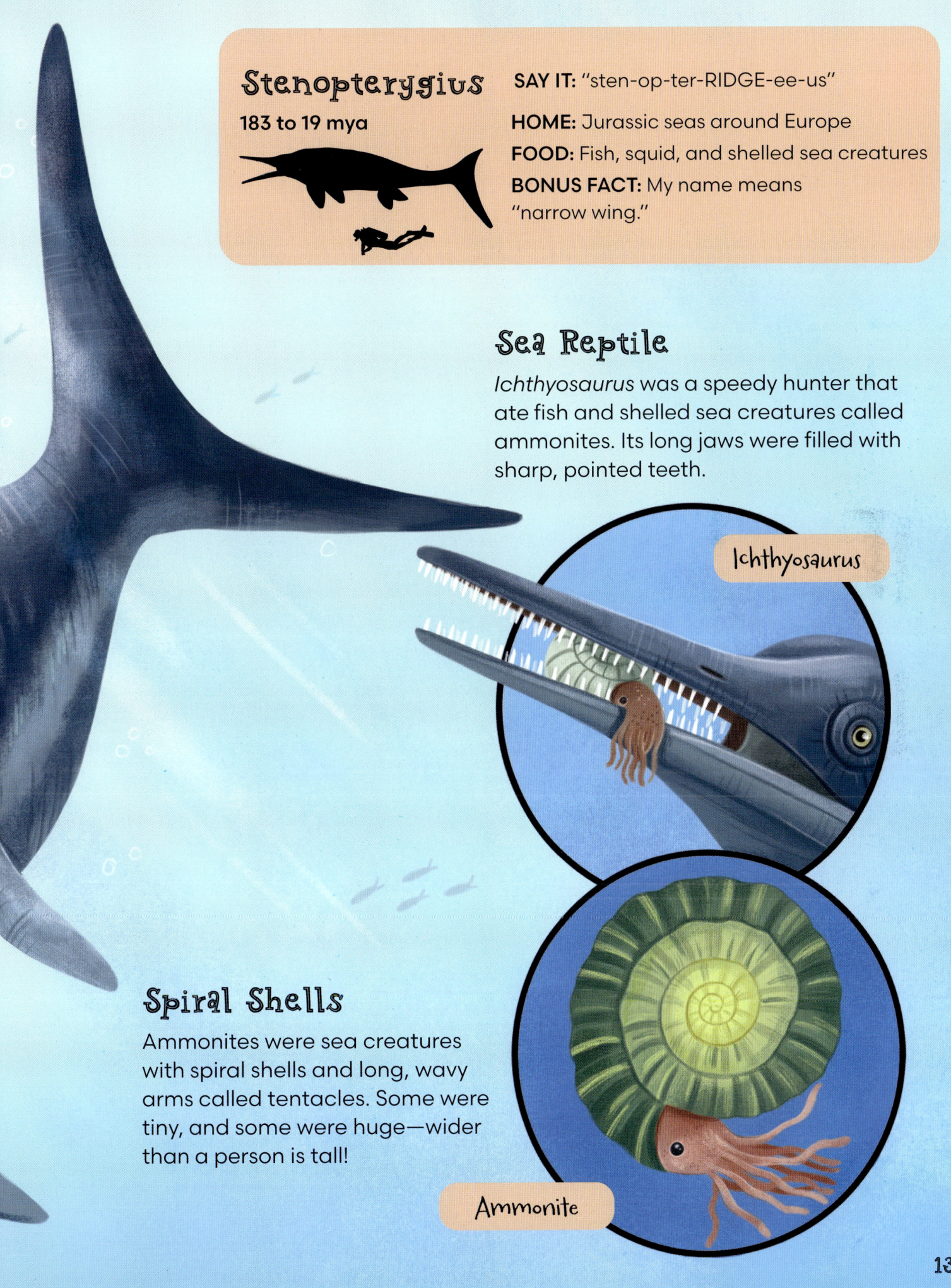

Stenopterygius
183 to 19 mya

SAY IT: "sten-op-ter-RIDGE-ee-us"
HOME: Jurassic seas around Europe
FOOD: Fish, squid, and shelled sea creatures
BONUS FACT: My name means "narrow wing."

Sea Reptile

Ichthyosaurus was a speedy hunter that ate fish and shelled sea creatures called ammonites. Its long jaws were filled with sharp, pointed teeth.

Ichthyosaurus

Spiral Shells

Ammonites were sea creatures with spiral shells and long, wavy arms called tentacles. Some were tiny, and some were huge—wider than a person is tall!

Ammonite

Fierce Hunters

Dinosaurs lived alongside lots of other fierce animals. Meat-eating dinosaurs both big and small roamed the land, and huge crocodiles lurked in rivers and lakes. In the oceans, giant reptiles called mosasaurs hunted fish, giant squid, and turtles.

Mighty Mosasaurs

Swimming reptiles called mosasaurs grew to huge sizes. *Tylosaurus* had giant jaws filled with rows of sharp teeth. It probably ate anything it came across, from fish, to turtles, to seabirds. Like all reptiles, *Tylosaurus* needed to breathe air, so it would come to the surface to do this. If it spotted a tasty pterosaur or bird, it would snap them up!

Tylosaurus

Megalosaurus

Mega Dino

Megalosaurus was a powerful hunter. This dinosaur was the first to be given a proper name by dinosaur scientists.

Monster Croc

Crocodiles have lived on Earth since before dinosaur times—and they are still alive today. The prehistoric monster *Deinosuchus* is the biggest ever to have lived.

Deinosuchus

Tylosaurus

90 to 66 mya

SAY IT: "ty-low-SORE-rus"

HOME: Cretaceous oceans around North America

FOOD: Fish, swimming reptiles, flying reptiles, and seabirds

BONUS FACT: I am closely related to lizards and snakes that are alive today.

Fliers and Gliders

During the Age of Dinosaurs, reptiles called pterosaurs flew through the skies. Some were big and some were small, but they were all good fliers. Over time, smaller, feathered dinosaurs adapted to gliding through the air. Eventually, some dinosaurs became more birdlike, and their bodies adapted for flight. The birds that we see today are living dinosaurs.

Microraptor

Micro Glider

Tiny *Microraptor* was a feathered dinosaur that could probably glide. It may have held its arms and legs out wide to pounce on prey below it.

Tupandactylus
12 mya

SAY IT: "too-pan-DACK-tih-lus"

HOME: Cretaceous woods around Brazil in South America

FOOD: Fish and small animals on land, and possibly fruit, too

BONUS FACT: My wing bones were hollow, like a bird's.

Crested Soarer

Pterosaurs took to the sky on wings covered by a thin layer of skin. Their bodies were very light. Some pterosaurs had strange-looking crests on their head. These crests may have helped them find a mate. *Tupandactylus* had a tall, sloping crest and a hard, beaky mouth. It may have soared over the ocean to scoop up fish to eat or hunted small animals on land.

Tupandactylus

Confuciusornis

Taking to the Air

Confuciusornis was a prehistoric bird. It had a long, trailing tail and could probably fly or glide short distances.

Dinosaurs Rule the Land

Dinosaurs were a successful animal group. There were many different kinds, from feathered carnivores to sharp-horned giants that browsed on leaves. Most plant-eating dinosaurs probably lived in herds. But these herds would have attracted hungry hunters on the lookout for their next meal.

Crunching Teeth

Albertosaurus was a fierce dinosaur with long teeth and strong jaws for crunching bones. It may have hunted in packs to attack big, plant-eating dinosaurs.

Albertosaurus

Saltasaurus

Bumpy Skin

Many plant-eating dinosaurs, such as *Saltasaurus*, had bony lumps and bumps on their skin to protect them from the teeth of hungry hunters.

Parasaurolophus

76 to 73 mya

SAY IT: "PA-ra-sore-OL-off-us"

HOME: Cretaceous plains, forests, and swamps of North America

FOOD: Leaves, twigs, ferns, pine needles

BONUS FACT: Some scientists once thought my head crest was a snorkel!

Parasaurolophus

Herd Life

The weird head crest of *Parasaurolophus* may have helped it make trumpeting calls or attract mates. This big dinosaur was a peaceful plant-eater that lived in family groups. Living as a group, or herd, was the best way to stay safe from danger and protect young dinosaurs.

How It Ended

About 66 million years ago, a giant rock from space smashed into Earth. This rock, or asteroid, killed many animals instantly. Giant waves from the oceans swept onto land, and volcanoes erupted. Smoke and ash made the air hard to breathe and blocked out light from the Sun.

Death of Dinosaurs

Without enough light from the Sun, the Earth began to cool. Fewer plants could grow, and this meant that plant-eating dinosaurs, such as mighty *Triceratops*, had less food. As the plant-eaters died, the meat-eating dinosaurs had no food. Slowly, the dinosaurs went extinct, or died out. It wasn't just the dinosaurs that died. Pterosaurs, mosasaurs, and lots of other animals went extinct, too.

Smashing into Earth

The asteroid that hit Earth made a massive crater. Dust, fumes, and fires began to spread. Much of life on Earth was wiped out, including the dinosaurs.

Triceratops

68 to 66 mya

SAY IT: "try-SEH-ra-tops"

HOME: Cretaceous forests of North America

FOOD: Leaves, twigs, roots, fruit, seeds, ferns, palms

BONUS FACT: I was one of the last dinosaurs to appear on Earth.

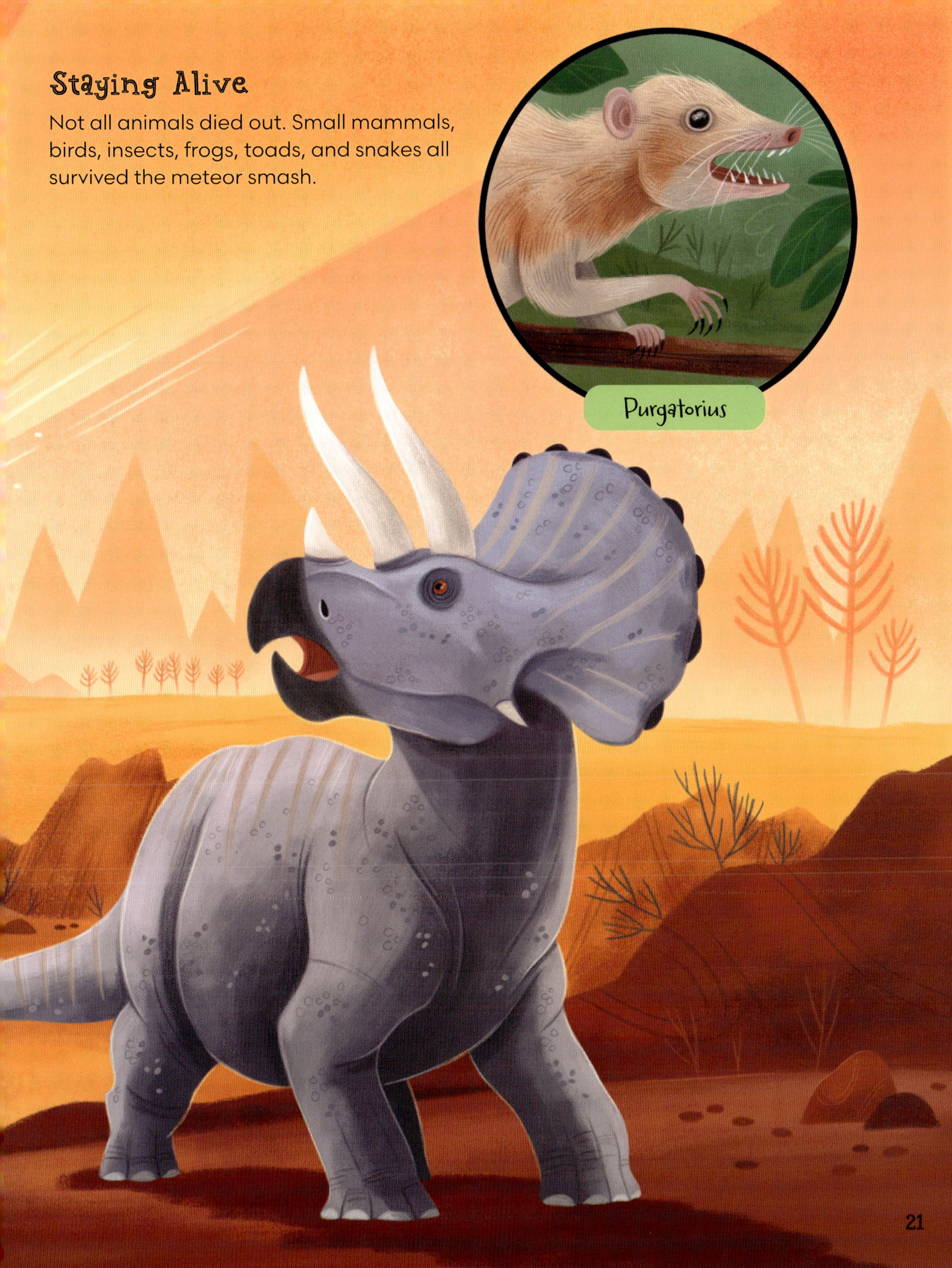

Staying Alive

Not all animals died out. Small mammals, birds, insects, frogs, toads, and snakes all survived the meteor smash.

Purgatorius

Rise of the Mammals

The first mammals were small and lived alongside the dinosaurs. After the asteroid smash, warm-blooded mammals adapted to new conditions on Earth. They began to grow bigger. As Earth warmed up, giant sloths, apes, and elephants roamed the land. Later, when Earth was gripped in an ice age, herds of huge, long-haired mammoths lived on snowy plains.

Tusks and Fur

Woolly mammoths were a type of prehistoric elephant. These huge animals had long, waterproof fur to keep them warm and big, curved tusks. Mammoths lived in herds, like elephants do today, and probably spent all their time eating grass that grew under the snow and ice. Mammoth ears were small to keep too much body heat from escaping.

Cave Bear

Smilodon

Cave Bear

Huge bears made their homes in caves. The cave bear was about the size of today's polar bear. It probably ate plants, fruits, and some meat.

Mighty Fangs

Smilodon was a huge prehistoric cat with long fangs. It could open its jaws wide to sink its teeth into prey.

Woolly Mammoth

SAY IT: "Wul-ee MAM-oth"

SIZE:

HOME: Frozen grasslands of North America, Europe, and Asia

FOOD: Mainly grass, and sometimes leaves

BONUS FACT: I had a fatty hump on my back that gave me energy if food was scarce.

How We Know

Scientists called paleontologists (pay-LEE-on-TOL-o-jists) study prehistoric life. The main way they do this is by looking at fossils. When an animal dies, its remains may be preserved if conditions are right. Over millions of years, the animal's body parts are replaced by stone. Hard parts such as bones, teeth, and claws make good fossils. Very occasionally, remains of soft parts become fossilized, too.

Looking at Fossils

Fossils are usually found when rock is worn away. Scientists dig them out of the ground with tools and take them back to a laboratory to study them carefully.

Borealopelta

Spiky Plant-Eater

Borealopelta was a spiky, plant-eating dinosaur. Fossil remains have been found of one that was swept out to sea after it died. It was buried on the seabed, and its fossils even show detail of its skin.

Borealopelta

110 mya

SAY IT: "BOH-ree-AH-loh-PEL-tuh"

HOME: Cretaceous forests of North America

FOOD: Ferns and leaves

BONUS FACT: My fossils are so detailed that even the red-brown of my skin is preserved.

How Fossils Form

Fossils form if an animal, such as a dinosaur, dies under the right conditions. The body is covered by mud or silt, and the soft parts usually rot away. More muddy layers build up on the amimal's remains, and over millions of years, they turn to rock. Earth's rocky layers are constantly pushing against each other. If the rock pushes up to the surface, the weather may wear it away to reveal the fossil.

Peaceful Plant-Eaters

Plants were the main food for most kinds of dinosaurs. Plant-eaters, or herbivores, feasted on leaves, ferns, and tough vegetation. Some of these dinosaurs grew to huge sizes and spent all their time eating to provide energy for their giant bodies. Others had bony plates, sharp horns, and clubbed tails to protect them from hungry predators.

Useful Hands

Iguanodon had a big, spiky thumb claw on each hand. The three middle fingers were joined together. The smallest finger could grip food for *Iguanodon* to eat.

Sharp Mouth Gills

Iguanodon had a hard "beak" to help it graze. Grinding teeth in its jaws mashed up tough leaves and trees.

Iguanodon

Spiky Plant-Eater

Although it walked on four legs, *Iguanodon* could stand on its back legs to reach food higher up. Its spiky thumb claws would have been useful for holding food or even for jabbing hungry meat-eaters. *Iguanodon* probably lived in herds. It belonged to a group of dinosaurs called ornithopods. These big plant-eaters had birdlike hips and walked on four legs.

Iguanodon
126 to 122 mya

SAY IT: "ig-WAH-noh-don"

HOME: Cretaceous woodlands of Europe, Africa, Asia, and North America

FOOD: Leaves, ferns, and flowering plants

BONUS FACT: I was the second dinosaur to be named by dinosaur scientists!

Altirhinus: Duck-Billed Dinosaurs

Altirhinus was a big dinosaur with a flat, beaky mouth and a bulge on its snout. It belonged to a group of dinosaurs called hadrosaurs, also known as duckbills. It had a stiff tail to help it balance.

Beaks and Crests

All hadrosaurs had flat, beaky mouths and batteries (rows) of teeth. Some had amazing crests on their head, which may have helped them make loud calls or trumpets. These dinosaurs probably lived in herds and journeyed long distances to find fresh food and safe areas to lay their eggs.

Altirhinus

Caring Mother

Maiasaura was another type of hadrosaur. Like other duck-billed dinosaurs, it cared for its young and may even have brought food to the nest.

Maiasaura

Big Nose

Perhaps *Altirhinus* used its big snout to call out to other dinosaurs. Or maybe it acted as a display to attract a mate. We may never know!

Altirhinus

Altirhinus

107 to 100 mya

SAY IT: "al-tee-RY-nus"

HOME: Cretaceous woodlands and plains in Asia

FOOD: Low-growing plants, ferns, leaves, and twigs

BONUS FACT: My name means "high nose"!

Stegosaurus: Spikes and Plates

Some dinosaurs were spiky! *Stegosaurus* also had a double row of tall, bony plates on its back. The plates may have helped *Stegosaurus* to attract mates. Because *Stegosaurus* was big and slow-moving, it needed some way of defending itself. At the end of its tail were four bony spikes that could be used as a weapon against attackers.

Stegosaurus

Danger Tail

The bony spikes at the end of *Stegosaurus*'s tail are called a thagomizer. They could cause a nasty injury to another dinosaur.

Tiny Brain

Stegosaurus had a small head even though its body was big. This means that it had a tiny brain!

Were Dinosaurs Clever?

Big dinosaurs like *Stegosaurus* had small brains compared to their big bodies, which meant that they were not very smart. But other dinosaurs may have been more intelligent. Small meat-eaters like *Troodon* and *Compsognathus* may have had bigger brains for their body size. They had good eyesight and senses of smell, and they could run fast.

Stegosaurus

155 to 145 mya

SAY IT: "STEG-oh-SORE-us"

HOME: Jurassic woods in Europe and North America

FOOD: Ferns, moss, leaves, conifers, and fruits

BONUS FACT: I could use my spiky tail to fight off fierce *Allosaurus*!

Ankylosaurus: Tough Skin

Some dinosaurs were like tanks! *Ankylosaurus* had hard, bony lumps all over its body. These lumps helped to protect *Ankylosaurus* from meat-eating predators. A bony club at the end of its tail made a good weapon. *Ankylosaurus* could swing its tail to smash into any dinosaur that came too close.

Ankylosaurus

Body Weapons

Ankylosaurs used their size and weapons to stay safe. Their skin was covered in so many lumps and plates, that they would have been very difficult to attack. These lumps are also found on the bodies of reptiles today, such as crocodiles and turtles. *Ankylosaurus* also had horns on the side of its head to protect its skull.

Ankylosaurus

68 to 66 mya

SAY IT: "an-KIH-loh-SORE-us"

HOME: Cretaceous forests of North America

FOOD: Ferns, fruits, and shrubs

BONUS FACT: My mighty tail club could break the bones of an attacker.

Hard Head

Even the head and face of *Ankylosaurus* were covered in plates of bone, including its eyelids!

Edmontonia

Sharp Spikes

Edmontonia had long shoulder spikes that stuck out to the sides. These scary spikes meant that this dinosaur would have been very hard to attack.

Psittacosaurus: Caring for Young

Because dinosaurs lived so long ago, we can only find out about them from their fossils. One fossil discovery of *Psittacosaurus* reveals that this dinosaur may have looked after its babies in a group, a little bit like a nursery. Caring for young like this would have given them more protection from predators.

Looking after Young

Perhaps *Psittacosaurus* shared babysitting duties with family members. Some kinds of birds do this today. We know that some dinosaurs did care for their young by building nests and even sitting on eggs to keep them warm. When baby *Psittacosaurus* hatched, they walked on four legs, but as they grew bigger, they were able to walk upright on their longer hind legs.

Dinosaur Egg

All dinosaurs laid eggs, just as most reptiles do today. Inside its egg, a baby dinosaur had a supply of food from a yolk.

Breaking Out

It may have taken a few months before a baby dinosaur was big enough to break out of its egg.

Kosmoceratops: Horns and Frills

One group of dinosaurs had long, sharp horns and frills of bone around their necks. They were the ceratopsians. Most ceratopsians were huge. Their giant bodies were supported by four stocky legs, and their enormous skulls carried the frill of bone and long horns. These giants would have been a match for fierce hunting dinosaurs.

Styracosaurus

Horned Face

Styracosaurus had a tall, pointed horn on its nose and several more on its neck frill. It probably ate low-growing plants and lived in herds.

Chasmosaurus

Swept-Back Frill

The neck frill of *Chasmosaurus* was huge! The horns on its face were smaller than those of some other ceratopsian dinosaurs.

Kosmoceratops
76 to 75 mya

SAY IT: "Koz-mo-SEH-ra-tops"

HOME: Cretaceous forests of North America

FOOD: Low-growing plants, ferns, fruits, and leaves

BONUS FACT: I had up to ten spikes on my neck frill.

Bony Necks

Kosmoceratops had a giant head and a wide, bony neck frill. Three sharp horns on its head may have helped to protect it against meat-eating dinosaurs. The neck frill may have been used to attract mates and helped *Kosmoceratops* recognize others of its kind. Although this dinosaur looked scary, it browsed on plants that it sliced up with its sharp, beak-like mouth.

Kosmoceratops

Torosaurus: Mighty Skulls

Fossils of *Torosaurus* show that it had a massive skull—one of the biggest of any land animal that has ever lived. This ceratopsian dinosaur had a huge, bulky body. It may have moved around in herds to keep younger dinosaurs safe from predators and to look for fresh food.

Snapping Beaks

Ceratopsians had beaky mouths for snapping at plants. Their beaks were made of keratin, which is what human nails and hair are made of. There have been more than 60 kinds of ceratopsian dinosaurs discovered so far.

Torosaurus

Torosaurus
68 to 66 mya

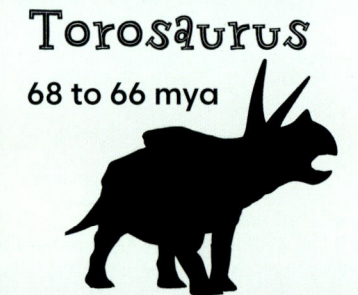

SAY IT: "tor-oh-SORE-us"

HOME: Cretaceous forests of North America

FOOD: Ferns, fruits, and shrubs

BONUS FACT: Although my neck frill was big, it wasn't very strong.

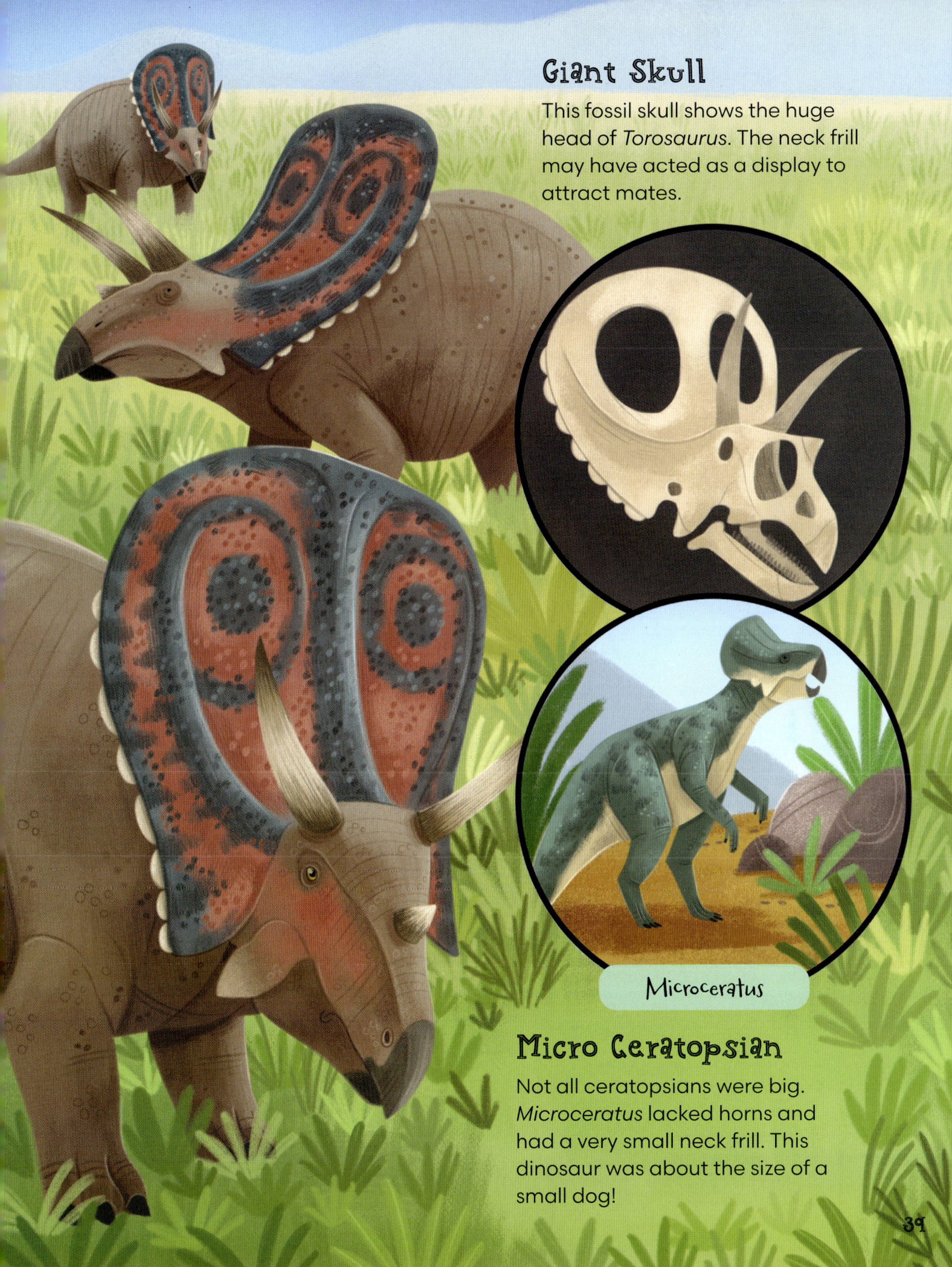

Giant Skull

This fossil skull shows the huge head of *Torosaurus*. The neck frill may have acted as a display to attract mates.

Microceratus

Micro Ceratopsian

Not all ceratopsians were big. *Microceratus* lacked horns and had a very small neck frill. This dinosaur was about the size of a small dog!

Pachycephalosaurus: Bone-Heads

Pachycephalosaurus was a medium-sized dinosaur. It had a thick shield of bone covering the top of its head, and it could move quickly on two legs. The bony head may have acted as a protective helmet because *Pachycephalosaurus* used its head in fighting competitions for mates.

Super Skull

This fossil shows the thick skull of *Pachycephalosaurus*. Large eye sockets mean that this dinosaur probably had excellent vision.

Dome Head

Stegoceras was a small dinosaur in the same group as *Pachycephalosaurus*. It also had a thick covering of bone on its head, giving it a dome-shaped appearance.

Stegoceras

Pachycephalosaurus

SAY IT: "pak-ee-SEF-ah-lo-SORE-us"

70 to 66 mya

HOME: Cretaceous woods and forests of North America

FOOD: Mainly plants but possibly small insects and lizards, too

BONUS FACT: My thick skull was more than 20 cm (9 inches) thick!

Pachycephalosaurus

Head to Flank

Although its skull was superthick, some dinosaur scientists think that *Pachycephalosaurus* may have tried butting into each other's sides, or flanks, instead of butting heads. This would have been a lot safer than smashing skulls.

Ingentia: Getting Bigger

Over millions of years, some plant-eating dinosaurs began to get bigger. They had longer necks that helped them reach food high up in trees. Their bigger bodies helped to protect them from fierce hunters. These plant-eaters began to spread around the world.

Early Giant

Huge *Ingentia* lived at the end of the Triassic Period, when most dinosaurs were small. Fossils show that *Ingentia* had air spaces along its neck and lungs that may have helped keep it cool—and meant that it didn't weigh too much. A bigger digestive system allowed *Ingentia* to eat more, and it probably kept growing throughout its life.

Plateosaurus

Before the Giants

Plateosaurus lived before the true giant dinosaurs. It had a long neck and tail, and was a big dinosaur for its time.

Ingentia
210 to 205 mya

SAY IT: "in-JEN-tee-uh"

HOME: Triassic forests of South America
FOOD: Leaves, branches, and ferns
BONUS FACT: I weighed as much as two elephants!

Ingentia

Air sacs made breathing more efficient.

Lighter bones

Light Bones

How did heavy dinosaurs move? Fossils show that *Ingentia* had hollow spaces, or air sacs, in its bones. This meant that its skeleton was light, helping it to move more easily.

Diplodocus: Jurassic Giants

Sauropods were the biggest dinosaurs ever to walk the Earth. They had long necks and tails and huge, barrel-shaped bodies. Their small heads were filled with peg-like teeth, and they walked on pillar-like legs. Sauropods could reach leaves in the tallest trees as well as graze on low-growing plants.

Peg Teeth

Diplodocus had long, blunt, peg-like teeth. These were good for raking leaves from plants.

Lizard Feet

Giant dinosaurs like *Diplodocus* had massive feet with sharp claws. Small dinosaurs would have needed to stay out of the way of these big-footed monsters.

Super Sauropods

Sauropods were able to grow so big because there were so many plants for them to feast on. Their size kept meat-eating dinosaurs away, so sauropods had little to fear. *Diplodocus* is one of the best-known sauropod dinosaurs. It had a superlong tail that it may have used like a whip to warn off hunters if they came too close.

Diplodocus

Diplodocus
154 to 152 mya

SAY IT: "dip-LOH-doh-kus"

HOME: Jurassic woods, forests, and plains of North America

FOOD: Plants, trees, leaves, ferns, pine cones, and needles

BONUS FACT: My skeleton has been in lots of museum displays around the world!

Sauroposeidon: Supersized!

Sauropod dinosaurs were enormous. But some were so big, they may have made the ground shake as they walked! *Sauroposeidon* was once thought to be the biggest dinosaur ever. It browsed on leaves from treetops, reaching food high up that other dinosaurs could not get to.

Working out Size

Sauroposeidon was a true giant. But was it the biggest dinosaur? Fossil bones help scientists figure out the sizes of dinosaurs. Although *Sauroposeidon* was enormous, it wasn't the biggest dinosaur. Its body was fairly lightly built, but this dinosaur still weighed as much as ten African elephants!

Fierce Hunter

Acrocanthosaurus lived at the same time as *Sauroposeidon*. It may have hunted young sauropods before they grew too big.

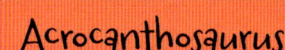
Acrocanthosaurus

Gastroliths

Plant-eating dinosaurs such as sauropods ate lots of tough vegetation. They may have swallowed stones called gastroliths to help mash up food in their stomachs.

Sauroposeidon

Sauroposeidon
112 mya

SAY IT: "SORE-oh-puh-SY-don"

HOME: Around swamps and plains in North America

FOOD: Leaves, ferns, palms, and flowering plants

BONUS FACT: My name means "earthquake lizard."

Argentinosaurus: Last of the Giants

Towards the end of the Age of Dinosaurs, some sauropod dinosaurs became even bigger. These were the titanosaurs. They had the same long necks, small heads, and whiplike tails, but their bodies were usually bigger and bulkier than those of other sauropods. Titanosaurs lived all over the world.

Strong Body

Argentinosaurus was one of the biggest dinosaurs ever. It had strong, straight legs and wide hips, which allowed it to move its giant body more easily. Because it was so big, it didn't have to worry about being attacked by hunting dinosaurs. *Argentinosaurus* may have lived in herds and journeyed for long distances to look for fresh food to eat.

Fossil Footprints

Fossils also form from footprints if the conditions are right. These fossils show us just how big titanosaurs were!

Long Neck

Argentinosaurus had a superlong neck. It probably ate food close to the ground as well as up high. Holding its long neck up all the time would have been hard work!

Argentinosaurus

SAY IT: "ar-jen-TEE-no-SORE-uss"

96 to 92 mya

HOME: Cretaceous woods, plains, forests of South America

FOOD: Plants, leaves, trees, ferns, fruit

BONUS FACT: My feet were padded, which helped me spread my weight.

Feathers, Beaks, and Claws

Not all dinosaurs were big meat-eaters or giant plant-eaters. Some were tiny—no bigger than a chicken. These smaller dinosaurs would look very strange to us. Many had feathers and looked almost birdlike. They ate plants, seeds, small reptiles, mammals—and other dinosaurs.

Fancy Feathers

The arms of *Caudipteryx* had long feathers. Its tail had a fan of feathers that may have been used as a display to show off to a mate like many birds do today.

Toothy Beak

Caudipteryx had a hard, beaky mouth to help it peck at plants and seeds. It also had teeth in its top jaw for snapping up insects and small lizards to eat.

Bird or Dinosaur?

Scientists have found many fossils of tiny *Caudipteryx*. These show that it had bird and reptile features. Its clawed arms looked like wings, but *Caudipteryx* could not fly. Its beaky mouth had reptile teeth. The tail was long and stiff like a dinosaur's but had a feathery fan. This dinosaur looked very birdlike. *Caudipteryx* was about the size of a turkey.

Caudipteryx

Caudipteryx
125 mya

SAY IT: "kaw-DIP-tuh-riks"

HOME: Cretaceous forests, rivers, and lakes in China

FOOD: Plants, seeds, insects, and lizards

Bonus fact: I only had teeth in my top jaw—they looked like pointed fangs!

Citipati: Caring for Eggs

All dinosaurs laid eggs, and fossil remains tell us that some dinosaurs cared for them, too. These dinosaurs sat on their eggs to keep them warm—and to keep them safe from other animals. This is called incubation, and most birds do this today.

Eggs in the Nest

Citipati laid its eggs in a nest made from earth. It used its body heat to keep the eggs warm and help the baby dinosaurs grow until they were ready to hatch.

Breaking Out

We don't know if *Citipati* cared for its young when they hatched. But it may have done so, because it took such good care of its eggs.

Citipati

Protecting Eggs

Citipati was a medium-sized dinosaur. It had a feathered body, long legs, and a beaky mouth. It certainly looked very birdlike. Fossils found of *Citipati* on its nest show that it may have been protecting its eggs from a sandstorm when it died. Other prehistoric animals, including dinosaurs, would also eat any eggs they came across.

Citipati

75 to 71 mya

SAY IT: "SIT-ee-PAT-ee"

HOME: Cretaceous deserts in Asia

FOOD: Plants, seeds, and insects

Bonus fact: It is thought that *Citipati* dads looked after the eggs!

Oviraptor: Fast and Feathered

Oviraptor was a small, feathered, birdlike dinosaur with a bony head crest. It could run fast on long legs and had three sharp-clawed toes on each foot. The beak of *Oviraptor* was strong and could probably crush tough nuts, seeds, and even shellfish. *Oviraptor* may have snacked on small mammals and lizards, too.

Gigantoraptor

Feathered Giant

Gigantoraptor was huge. It was bigger than the biggest bird alive today—the ostrich. This feathered dinosaur looked fierce, but it was a plant-eater.

Crushing Beak

Oviraptor's beak was superstrong. It was sharp, hard, hornlike, and covered in keratin—the same material that your hair and fingernails are made of.

Oviraptor

Why Were Some Dinosaurs Feathered?

Oviraptor, *Citipati*, and *Gigantoraptor* belonged to the same dinosaur group—the oviraptorosaurs. They were closely related, despite their differences in size. All had feathers, but these feathers weren't for flight—dinosaurs could not fly. It is most likely that the feathers were for warmth and to attract mates.

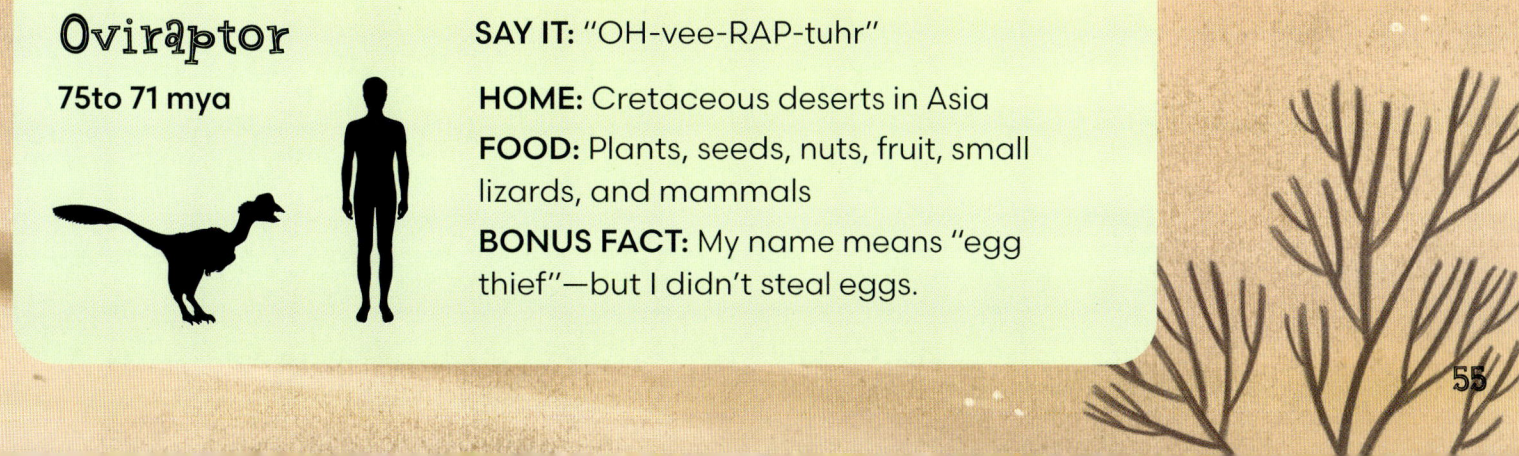

Oviraptor
75 to 71 mya

SAY IT: "OH-vee-RAP-tuhr"

HOME: Cretaceous deserts in Asia

FOOD: Plants, seeds, nuts, fruit, small lizards, and mammals

BONUS FACT: My name means "egg thief"—but I didn't steal eggs.

Deinocheirus: Giant Arms and Claws

Some of the strangest dinosaurs discovered have been named from just a few fossils. *Deinocheirus* was huge. It had superlong arms, a duck-billed mouth, and a hump on its back. It took scientists a long time to figure out what kind of dinosaur it was.

Long-Armed Giant

Deinocheirus belongs to a group known as "ostrich dinosaurs." Most of the dinosaurs in this group had feathers and could run fast like an ostrich does today. But *Deinocheirus* was too big to run at speed, although it most likely was feathered. It may have waded in rivers and lakes to pull up plants to eat with its long arms and claws.

Deinocheirus

Deinocheirus
70 mya

SAY IT: "DIE-no-KY-rus"

This is where I live: Around Cretaceous rivers and lakes in Asia

FOOD: Fish, amphibians, and plants

BONUS FACT: I swallowed stones to help mash up plants in my stomach!

Giant Arms

Deinocheirus had the longest arms of any dinosaur. For many years, these were the only fossils that existed of this dinosaur!

Therizinosaurus

Grabbing Claws

Therizinosaurus was another strange-looking dinosaur. Its claws were incredibly long, and probably used for grabbing plants and fruit to eat.

Compsognathus: Tiny Hunter

Some meat-eating dinosaurs were tiny. These small hunters were fast and light-footed and could chase lizards and other fast-moving prey into shrubs and undergrowth. Many had a covering of feathers to keep them warm and sharp teeth and claws to grab prey.

Sharp-Toothed Hunter

Compsognathus's jaws were filled with sharp teeth. It had good eyesight and could move fast to grab lizards and mammals. It had a long, stiff tail. This would have helped it balance as it ran, twisted, and turned while hunting prey.

Sinosauropteryx

Small and Feathered

Fossils of *Sinosauropteryx* reveal a tiny dinosaur with fluffy feathers. The tail had bands of dark- and light-brown feathers, and the body feathers were brown, too.

Juravenator

Night Hunter

Juravenator may have been nocturnal. This small dinosaur could likely see well in the dark and was active at night, hunting small creatures to eat.

Compsognathus

150 mya

SAY IT: "komp-sog-NAY-thus"

HOME: Jurassic forests and woodlands of Europe

FOOD: Small lizards, insects, and tiny mammals

BONUS FACT: I was about the size of a chicken!

Compsognathus

Archaeopteryx: Dino Bird

Was *Archaeopteryx* a bird or a dinosaur? It had features of both, with reptile teeth and claws on its wings. Today, *Archaeopteryx* is seen as an early relative of birds. It was covered in feathers and may have been able to fly or glide for short distances. It also may have climbed into trees to sleep at night.

Ancient Bird

Fossils of *Archaeopteryx* have given scientists lots of information. Its feathers may have been mainly black, with some white. The wing feathers were capable of flight, but this ancient dino bird could not fly great distances. It may have jumped from trees to catch passing prey in the air, such as insects, and then glided to the ground.

Teeth and Tail

Unlike birds today, *Archaeopteryx* had rows of small, sharp teeth in its jaws. It also had a long tail, like a dinosaur.

Archaeopteryx

150 to 148 mya

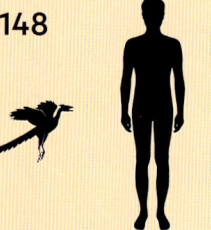

SAY IT: "AR-kee-OP-tuh-rix"

HOME: Jurassic wooded islands in Germany

FOOD: Insects and small reptiles

BONUS FACT: My name means "ancient wing."

Prehistoric Chick

The hoatzin is a living bird of South America. Hoatzin chicks are born with clawed wings to help them grip branches. They lose their claws as they get older.

Hoatzin

Archaeopteryx

Deinonychus: Killer Claws

Deinonychus belonged to a group of dinosaurs called dromaeosaurs (drom-EE-oh-SOREs). These dinosaurs are often referred to as raptors. Some were tiny, but others were big. All were fierce hunters with stabbing claws and sharp teeth. They ran fast, had good eyesight, and were smart, too. Some may have hunted in packs to attack bigger dinosaurs.

Terror Claw

Deinonychus had an extra-long claw on each foot. The claw could flick down to stab prey.

Dromaeosaurus

Powerful Jaws

Dromaeosaurus was a small dinosaur with a big head. Its jaws were strong and could probably crunch through bone.

Deinonychus

115 to 108 mya

SAY IT: "dy-NON-ik-uss"

HOME: Cretaceous swamps and tropical forests of North America

FOOD: Plant-eating dinosaurs and small mammals

BONUS FACT: My name means "terrible claw."

Feathered Killer

As well as long stabbing claws on each foot, *Deinonychus* had strong jaws and sharp teeth for biting prey. It could run fast and had a long bony tail that helped it balance and turn quickly. *Deinonychus* may have hunted in groups to take on larger dinosaurs. A pack of these feathered killers would have been a scary sight!

Deinonychus

Velociraptor: Small and Deadly

Velociraptor was a small, fierce dinosaur. It had a long slicing claw on each foot and could run fast. Feathers covered its body. This speedy hunter chased prey that it grabbed and stabbed with sharp claws. Like many other dinosaurs, it was probably warm blooded.

Narrow Jaws

Velociraptor had small teeth that curved backward. It would have used its claws to attack prey, then bitten off small chunks to eat.

Tiny Killer

Bambiraptor was the smallest of the dromaeosaurs. It was about the size of a small dog, and it had sharp claws and snapping jaws for catching small lizards and amphibians.

Bambiraptor

Velociraptor

75 to 71 mya

SAY IT: "veh-LOSS-ee-rap-tuhr"

HOME: Cretaceous deserts of Asia

FOOD: Amphibians, lizards, and dinosaurs

BONUS FACT: I was also a scavenger and ate the remains of dead animals.

Velociraptor

Night Hunter

Fossils of *Velociraptor* show that it had a good sense of smell and big eyes. It may have hunted at night. Its feathers would have helped to keep it warm as temperatures dropped. One fossil discovery shows a *Velociraptor* locked in battle with a plant-eating *Protoceratops*.

Mighty Meat-Eaters

The biggest meat-eating (or carnivorous) dinosaurs walked on two strong legs, but many had small arms. Their bodies were muscular, and their heads contained powerful jaws and long, sharp teeth. These hunters preyed on plant-eating dinosaurs and the remains of dead animals.

Crested Killer

Dilophosaurus was a fierce predator with a weird bony crest on its head. It had long, sharp claws to grab its prey.

Antarctic Dinosaur

Cryolophosaurus lived in the Antarctic when forests grew along the coastline. The crest on its head may have helped *Cryolophosaurus* attract a mate.

Big for Its Time

Dilophosaurus was the top hunter of its time. It lived in the early part of the Jurassic Period when most dinosaurs were still small. Like *Cryolophosaurus* it had a head decoration—but it was even more of a show-off, with two crests! As the biggest hunter, it had little to fear and could move fast on strong legs. Its jaws were packed with sharp teeth.

Dilophosaurus
193 mya

SAY IT: "dy-LOFF-oh-SORE-us"

HOME: Jurassic swamps of North America

FOOD: Plant-eating dinosaurs, mammals, and the remains of dead animals

BONUS FACT: Some of my bones had air spaces—this made me light and fast for my size.

Ceratosaurus: Horned Hunter

Ceratosaurus was a scary sight. It had a huge head and hornlike ridges above its eyes—and a horn on its nose. This nose horn would have grown bigger as *Ceratosaurus* became an adult and may have helped it attract a mate. *Ceratosaurus* lived at the same time as bigger meat-eaters such as *Allosaurus*.

Bony Back

Ceratosaurus had a row of bony lumps running along its neck, back, and tail. These lumps, or scutes, would make it difficult for another dinosaur to take a bite out of *Ceratosaurus*. Other big carnivores that lived at the same time were *Allosaurus* and *Torvosaurus*. Perhaps this is why *Ceratosaurus* needed protection.

Ceratosaurus

Big Head

Ceratosaurus had a big head for its body size. But its jaws were not strong enough to crunch through bone.

Slicing Teeth

Long, slicing teeth helped *Ceratosaurus* bite and slash at prey. It may even have hunted fish in rivers and swamps.

Ceratosaurus
153 to 158 mya

SAY IT: "seh-RAT-oh-SORE-us"

HOME: Jurassic plains and swampy habitats

FOOD: Plant-eating dinosaurs, fish, crocodiles, and turtles

BONUS FACT: My my name means "horned lizard."

Allosaurus: Terror of the Jurassic

Allosaurus is one of the best-known dinosaurs from fossil remains. This fierce hunter was a top predator of its time, living alongside plant-eaters such as stegosaurs and giant sauropods. *Allosaurus* was big and powerful with sharp-clawed hands to help it grip and slash at prey.

Fossil Clues

Even compared to other Jurassic meat-eaters, *Allosaurus* was a giant. It probably hunted giant prey, too. Sauropod fossil bones have been found that show Allosaurus bite marks. Another fossil reveals that *Allosaurus* attacked mighty *Stegosaurus*—and was injured by the deadly thagomizer at the end of its tail.

Scavenging Remains

As well as hunting for food, *Allosaurus* would have eaten the remains of dead animals it came across. This is called scavenging.

Allosaurus
155 to 145 mya

SAY IT: "AL-oh-SORE-us"

HOME: Jurassic plains and forests of North America

FOOD: Plant-eating dinosaurs and the remains of other animals

Bonus fact: I grew new teeth throughout my life.

Allosaurus

Giant Jaws

Allosaurus could open its mouth wider than any other dinosaur. Gaping jaws helped it take big chunks out of prey.

Giganotosaurus: Big-Headed Monster

Some of the biggest, fiercest meat-eating dinosaurs lived toward the end of the Age of Dinosaurs. One of these was *Giganotosaurus*. It had a huge head and jaws packed with long, sharp teeth. This dinosaur may have hunted huge sauropods, taking bites out of them to injure them before closing in for the kill.

Giganotosaurus

Tiny Arms

Despite its huge body, the arms of *Giganotosaurus* were small and weak. Each hand had three clawed fingers

Saw Teeth

The long teeth of *Giganotosaurus* were serrated, with sharp, jagged edges like a saw. They were ideal for biting flesh.

Big Head

Giganotosaurus was one of the biggest meat-eating dinosaurs. It probably grew fast as a youngster and may have kept growing for most of its life. The huge head is the biggest of any known hunting dinosaur found so far. It was filled with air spaces to help keep it light.

Gigantosaurus
98 to 97 mya

SAY IT: "jig-an-OH-toe-SORE-us"

HOME: Cretaceous swamps and forests of South America

FOOD: Big, plant-eating dinosaurs and the remains of dead animals

BONUS FACT: I may have hunted in packs for giant sauropods.

Spinosaurus: Sail-Backed Killer

Spinosaurus is the biggest hunting dinosaur found so far. Unlike other big theropods, it had long jaws like a crocodile and a huge spiny sail on its back. *Spinosaurus* lived around rivers and lakes, hunting for fish as it used its long tail to push through water.

Hunting in Water

Fossils of *Spinosaurus* show that it was well-suited to life around water. Unlike most other dinosaurs, its bones were solid, with no air spaces. This may have helped it dive underwater or wade across swampy riverbeds. Its tail was long and flexible and shaped like a rudder to help push through water as *Spinosaurus* hunted for fish.

Snapping Jaws

The long jaws of *Spinosaurus* helped it snap up fish to eat. Its nostrils were set high up on the nose to allow *Spinosaurus* to breathe as it looked for fish at the surface.

Super Sail

The tall sail on the back of *Spinosaurus* has been a puzzle to scientists. It is possible that it was used as a display to attract mates.

Spinosaurus

112 to 93 mya

SAY IT: "spy-no-SORE-us"

HOME: Swamps of North Africa

FOOD: Fish, sharks, and other dinosaurs on land

BONUS FACT: My teeth were cone-shaped—perfect for spearing slippery fish!

Tyrannosaurus rex: King of the Dinosaurs

No other dinosaur is better known than *Tyrannosaurus rex*. This mighty carnivore lived at the end of the Age of Dinosaurs and was a dangerous hunter of big, plant-eating dinosaurs. Its jaws were strong enough to crunch through bone, and it had excellent senses of smell and vision. *T. rex* was the top hunter of its time.

Bone Cruncher

T. rex had the most powerful bite of any dinosaur. This bite force could shatter bone and the bony skin plates of big, well-protected dinosaurs such as *Ankylosaurus*. Because of its size, it would not have been able to run. Instead, it may have taken long strides to keep up with prey and deliver a killer bite. It may also have hunted at night to take prey by surprise

Using Senses

T. rex had forward-facing eyes and could easily spot prey. As well as hunting for food, *T. rex* used its powerful sense of smell to sniff out the remains of dead animals to eat.

Tyrannosaurus rex

68 to 66 mya

SAY IT: "ty-RAN-oh-SORE-us REX"

HOME: Cretaceous forests, plains, and swamps of North America

FOOD: Plant-eating dinosaurs and the remains of other animals

BONUS FACT: My arms were tiny but surprisingly strong!

Tyrannosaurus rex

Fossil Poop

Fossilized dinosaur poop is called coprolite. By examining the fossil poop of *T. rex*, scientists know that it ate hadrosaurs and huge ceratopsians such as *Triceratops*.

In Air and Water

During the Mesozoic Era, dinosaurs ruled the land. But they could not fly, and although some could likely swim, they did not live in water. Instead, dinosaurs shared their world with flying reptiles called pterosaurs. And rivers and oceans were home to fierce crocodiles and mighty sea reptiles.

Hunting for Fish

Pteranodon was a fish-eating pterosaur. It may have been able to dive beneath the waves to grab slippery fish with its long beak.

Incredible Crests

Many pterosaurs had head crests. These were probably used as a display to attract mates. *Tapejara* was a small pterosaur with a tall, thin head crest. This may have been green, blue, or red. We may never know for sure.

Tapejara

Pteranodon

Fantastic Fliers

Pterosaurs varied in size—some were tiny, and others were huge. But all could fly. They had light, hollow bones and strong wings covered by a thin layer of skin. *Pteranodon* flew over the ocean hunting fish. It may have lived in large groups like many seabirds do today. Like most reptiles, *Pteranodon* laid eggs on land, from which its young hatched.

Pteranodon
86 to 85 mya

SAY IT: "tuh-RAH-no-don"

HOME: Cretaceous seas and coasts of North America

FOOD: Fish

BONUS FACT: Unlike most pterosaurs, I was toothless.

Quetzalcoatlus: Giant of the Skies

One pterosaur was the size of a small plane—the mighty *Quetzalcoatlus*. It was the biggest flying animal ever, but its body was light. It flew through the skies looking for small dinosaurs to snack on. Despite its size, *Quetzalcoatlus* was likely a strong flier. It may have saved energy by soaring on warm air currents.

Taking to the Air

How did a pterosaur the size of a small plane take to the air? *Quetzalcoatlus* had powerful muscles in its back legs. It jumped high into the air from the ground and flapped its wings to get airborne. Once in the air, its light, hollow skeleton allowed *Quetzalcoatlus* to climb and soar on air currents.

Quetzalcoatlus

68 to 66 mya

SAY IT: "KWETS-ul-koh-AT-lus"

HOME: Cretaceous plains of North America

FOOD: Small dinosaurs, mammals, and reptiles, and it may have fed on insects, fruits, and plants.

BONUS FACT: I was 5 to 9 m (16 to 30 ft) long!

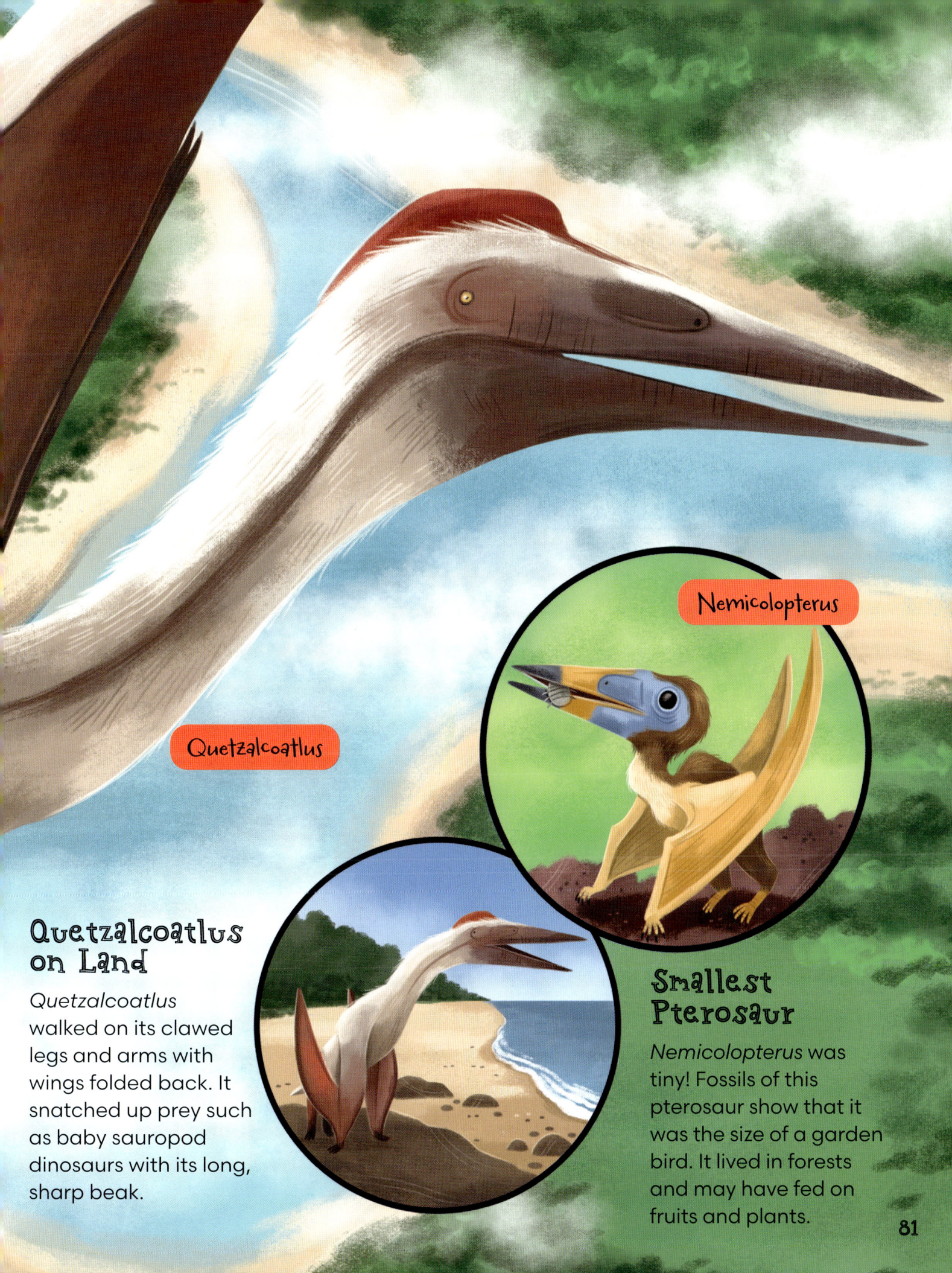

Quetzalcoatlus on Land

Quetzalcoatlus walked on its clawed legs and arms with wings folded back. It snatched up prey such as baby sauropod dinosaurs with its long, sharp beak.

Smallest Pterosaur

Nemicolopterus was tiny! Fossils of this pterosaur show that it was the size of a garden bird. It lived in forests and may have fed on fruits and plants.

Mosasaurus: Terror of the Seas

Some of the biggest, scariest reptiles lived in the ocean. *Mosasaurus* had a long, powerful body and tail, paddle-like flippers, and huge jaws like a crocodile. The jaws were packed with teeth to grab prey such as turtles, fish, sharks, and other sea creatures. This reptile wasn't fussy about what it ate.

Mosasaurus

Super Sight

Fossil skulls of *Mosasaurus* show that it had good vision. The eyes were big, helping *Mosasaurus* see well in dark water to take prey by surprise.

Turtle Killer

The jaws of *Mosasaurus* were strong enough to bite through the tough shells of turtles.

Swimming Machines

Mosasaurs were well adapted to living in the ocean. The long tail acted like a rudder to steer through the water. Like all reptiles, these fierce hunters had lungs and needed to come to the surface to breathe air. Mosasaurs may not have been fast swimmers, but they were powerful, so they could attack at high speed.

Mosasaurus
82 to 66 mya

SAY IT: "MOH-sah-SORE-us"

HOME: Cretaceous oceans across the world

FOOD: Turtles, fish, sharks, and other mosasaurs

BONUS FACT: I am related to modern-day snakes and lizards.

Psephoderma: Shellfish Crusher

Placodonts were a group of lizard-like reptiles that appeared in the Triassic Period. They lived along shallow coasts and ate shellfish, which they dug up from the seabed. Over time, placodonts developed shells to protect their bodies. *Psephoderma* had a beaky mouth and flat, blunt teeth to crush its food.

Weird Teeth

Placodus had weird teeth that stuck out and forward from its jaws. These teeth were useful for scraping up crunchy shellfish from the rocky seabed.

Placodus

Psephoderma

210 mya

SAY IT: "see-foe-DERM-uh"

HOME: Triassic coastal waters around Europe

FOOD: Shellfish

BONUS FACT: My double shell had a pebbly texture—my name means "pebbly skin."

Psephoderma

Spiky Shell

Cyamodus had a spiky double shell to protect it from predators. The shell also made *Cyamodus* heavier, so it could easily sink to the seabed to find shellfish to eat.

Cyamodus

Life in the Shallows

Placodonts like *Psephoderma* spent most of their time in shallow water. Like all marine reptiles, they needed to breathe air, so they may have spent time on land, too. It is not known if they laid their eggs on land or gave birth to young in the water. But they may have laid their eggs on beaches like turtles do today.

Archelon: Giant Turtle

The biggest living turtle today is the leatherback. During the Late Cretaceous Period, there lived a turtle that was twice as big—*Archelon*. This huge, shelled reptile swam slowly using four paddle-shaped flippers and feasted on fish, jellyfish, shellfish, and squid.

Baby Turtles

Archelon spent most of its life at sea. It returned to land to lay its eggs on beaches. When the baby turtles hatched, they may have been prey for small dinosaurs and seabirds. They needed to dash for the ocean waves to avoid being eaten.

Hesperornis

Cretoxyrhina

Tiny Wings

Hesperornis was a prehistoric seabird. It was flightless, with tiny wings and big feet. But it could swim and dive well to hunt for fish.

Monster Shark

Cretoxyrhina was a big, powerful prehistoric shark. Its body shape was very similar to today's great white shark. It feasted on giant turtles—and anything else that came along.

Archelon

80 to 74 mya

SAY IT: "ar-KEE-lon"

HOME: Cretaceous seas of North America

FOOD: Shellfish, fish, jellyfish, and squid

BONUS FACT: I had a hooked beak and strong jaws for crushing my food.

Archelon

Elasmosaurus: Monster of the Deep

Long-necked sea reptiles called plesiosaurs (pless-ee-oh SORES) inhabited prehistoric oceans. They had barrel-shaped bodies and strong flippers for swimming. Their heads were small, with jaws packed with long, sharp teeth for catching fish. One of the biggest plesiosaurs was *Elasmosaurus*.

Fanged Fish

Enchodus was a prehistoric fish with superlong fangs. It was a fierce hunter of smaller fish but was probably preyed upon itself by plesiosaurs.

Terror Teeth

The teeth of *Elasmosaurus* were long and thin. When it closed its jaws, the teeth overlapped each other. Any fish that it caught were speared and trapped.

Elasmosaurus
80 mya

SAY IT: "eh-LAZZ-moh-SORE-us"

HOME: Cretaceous seas around North America

FOOD: Fish, squid, and ammonites

BONUS FACT: I had 71 bones in my neck!

Longest Neck

All plesiosaurs had long necks, but *Elasmosaurus* had the longest neck known to date. This allowed it to easily pick off fish to eat from big shoals. *Elasmosaurus* could stay hidden from view, then reach its neck through the water, opening its jaws to snap up food.

Sarcosuchus: Giant Crocodile

Giant crocodiles lived in and around rivers and lakes during the Age of Dinosaurs. These fierce reptiles had huge jaws filled with teeth to grab their prey. Fossils of *Sarcosuchus* show that it is one the biggest crocodiles found so far. But did this giant feed on big prey, such as dinosaurs?

Sarcosuchus

Sarcosuchus
133 to 112 mya

SAY IT: "SAR-ko-SOO-kus"

HOME: Cretaceous rivers and lakes of Africa

FOOD: Fish, turtles, other reptiles, and the remains of dead animals

BONUS FACT: My tough skin was covered in thick, scaly lumps called scutes.

Getting Bigger

Prehistoric crocodiles kept growing throughout their lives. The long, narrow jaws of *Sarcosuchus* grew wider as it grew older and bigger. This meant that bigger prey, such as dinosaurs, may have been a source of food. Like modern crocodiles, *Sarcosuchus* hid below the surface to take prey by surprise.

Fish Lover

Sarcosuchus probably ate big fish such as *Mawsonia*, which were easier to catch. It may have feasted on turtles, too, and the remains of dead animals it came across.

Mawsonia

Nose Bulge

Sarcosuchus had a big bulge on its snout called a bulla. Scientists are unsure what this may have been for. It may have helped *Sarcosuchus* sniff out food.

Liopleurodon: Deep-Sea Hunter

Fierce sea reptiles called pliosaurs lived during the Jurassic and early Cretaceous periods. They had long, barrel-shaped bodies, short necks, and huge heads with jaws like a crocodile. Pliosaurs swam using powerful flippers. *Liopleurodon* was one of the biggest pliosaurs, and it could eat almost any animal it came across.

Super Fossil

In 2023, the fossil skull of a giant pliosaur was dug out of cliffs on a beach in Dorset, UK. The skull is one of the most detailed pliosaur fossils ever found.

Liopleurodon

Liopleurodon
166 to 140 mya

SAY IT: "LYE-oh-PLEUR-oh-don"

HOME: Jurassic seas of Europe

FOOD: Ammonites, remains of dead animals

BONUS FACT: I was closely related to the long-necked, small-headed plesiosaurs.

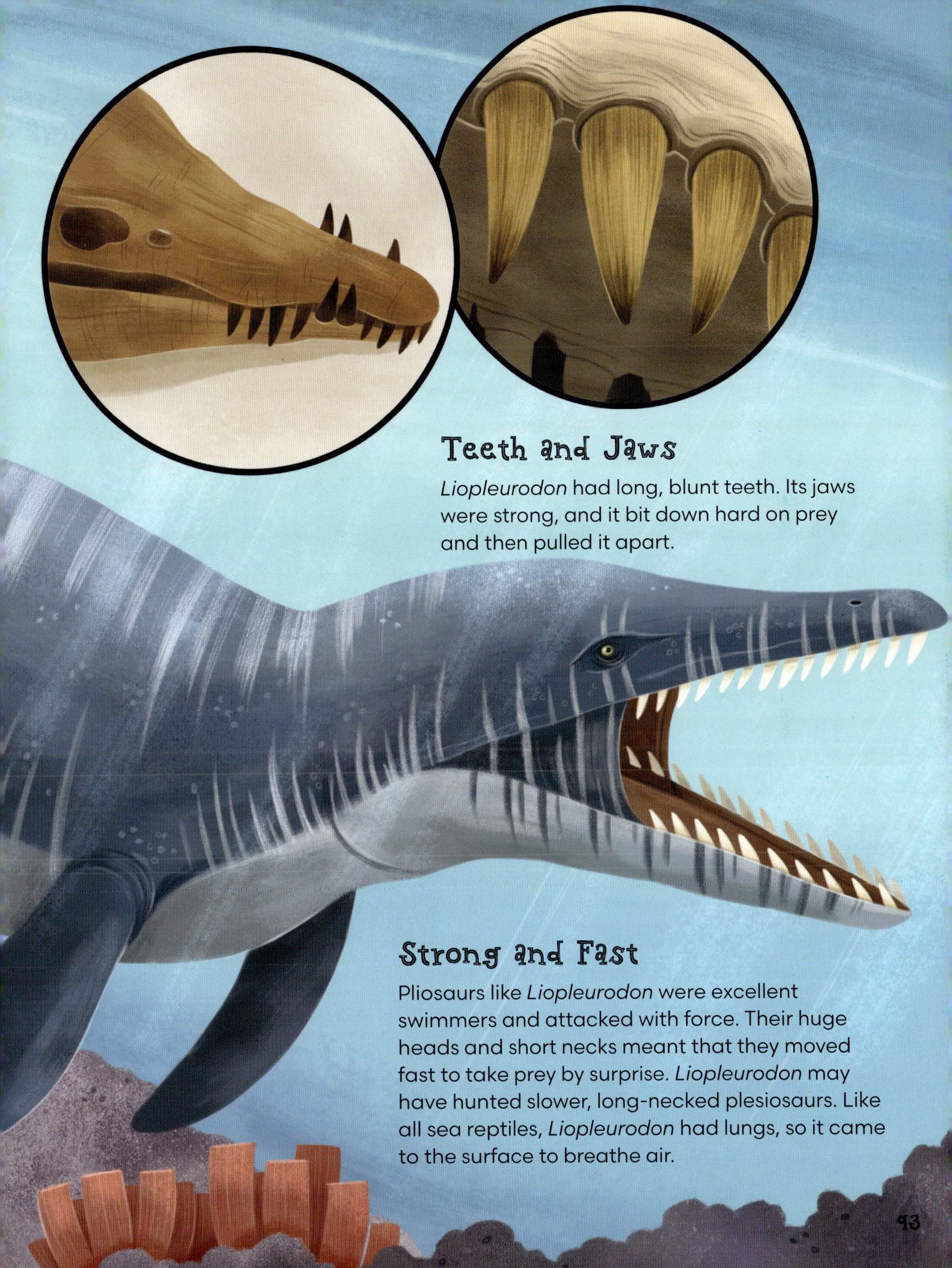

Teeth and Jaws

Liopleurodon had long, blunt teeth. Its jaws were strong, and it bit down hard on prey and then pulled it apart.

Strong and Fast

Pliosaurs like *Liopleurodon* were excellent swimmers and attacked with force. Their huge heads and short necks meant that they moved fast to take prey by surprise. *Liopleurodon* may have hunted slower, long-necked plesiosaurs. Like all sea reptiles, *Liopleurodon* had lungs, so it came to the surface to breathe air.

Glossary

ammonite
A shelled sea creature that lived in prehistoric oceans. Ammonites ranged in size and had wavy arms called tentacles to grab food from the water.

amphibian
Cold-blooded animals that lay their eggs in water. Frogs and toads are amphibians. *Diplocaulus* was a prehistoric amphibian.

batteries
Rows of teeth that some plant-eating dinosaurs had to help them chew and crush tough plants.

bulla
A bulge at the end of the snout of the prehistoric crocodile *Sarcosuchus*. This may have helped *Sarcosuchus* smell food or sense when animals were nearby.

carnivore
An animal that eats other animals.

coprolite
Fossil dinosaur poop.

crest
A thin piece of skin or bone on the heads of some dinosaurs and pterosaurs. Crests could be big or small and were probably used to attract mates. Some living birds and reptiles have crests.

energy
Animals need energy to live and move their bodies. Energy comes from the food that animals eat.

environment
The place, surroundings, and conditions in which animals and plants live. Some dinosaurs lived in a hot, dry environment.

evolve
When life evolves, it slowly changes to adapt to conditions such as weather and food supply. Dinosaurs evolved as Earth changed over millions of years.

extinct
When a type of animal dies out completely, it becomes extinct. Those animals no longer exist.

fish
Cold-blooded animals that live and breathe in water, such as sharks.

flippers
Body parts that help some sea creatures to swim. Mosasaurs and plesiosaurs had flippers.

fossil
The remains of a dead animal from long ago, such as a dinosaur. The remains have turned to stone.

gills
Body parts on fish and some kinds of amphibians. Gills help these animals breathe in water.

grassland
A large area of land in which lots of different types of grasses grow. Grass first grew toward the end of dinosaur times.

habitat
The place where an animal lives. A habitat can be in water, in a forest, or in a desert.

herbivore
An animal that feeds only on plants. Most dinosaurs were herbivores.

ichthyosaur
A prehistoric sea reptile that looked like a dolphin. Ichthyosaurs had flippers and long jaws filled with teeth.

incubation
When egg-laying animals such as birds sit on their eggs to keep them warm and safe. Some dinosaurs, such as *Citipati*, did this, too.

laboratory
A building where scientists use equipment to study materials and conduct experiments. Dinosaur fossils are studied in a laboratory.

mammal
A warm-blooded animal that gives birth to live young and feeds them on milk.

nocturnal
If an animal is more active at night than during the day, it is nocturnal. Some dinosaurs may have been nocturnal.

plains
Large, flat areas of land that can be hot, cold, wet, or dry. Many animals live on plains.

plesiosaur
A type of prehistoric sea reptile with a long neck, small head, and sharp teeth. Plesiosaurs had barrel-shaped bodies and four flippers.

pliosaur
A type of fierce prehistoric sea reptile related to plesiosaurs. Pliosaurs had short necks and long jaws like a crocodile.

predator
An animal that hunts and eats other animals. *Tyrannosaurus rex* was a predator.

prehistoric
The time before history was recorded or written down. Dinosaurs are prehistoric animals.

prey
An animal that is hunted for food by another animal. Plant-eating dinosaurs were prey for meat-eating dinosaurs.

pterosaur
A flying prehistoric reptile.

raptor
Dinosaurs in the dromaeosaur group are often referred to as raptors.

reptile
A cold-blooded animal with scaly skin. Most reptiles lay eggs. Dinosaurs were reptiles, although some may have been warm-blooded, and many had feathers.

sauropod
A huge, plant-eating dinosaur with a long neck, small head, and pillar-like legs.

scavenger
An animal that eats dead animals. Some meat-eating dinosaurs may have been scavengers.

scutes
Thick, bony lumps on the skin of some reptiles such as crocodiles. Many dinosaurs had scutes.

serrated
The jagged, sawlike edge on the teeth of some dinosaurs and prehistoric sea reptiles.

species
The different kinds of animals that make up a group. More than 60 species of ceratopsian dinosaurs have been discovered.

swamp
A large, wet area of land with rivers or lakes.

thagomizer
The four sharp bones at the end of *Stegosaurus*'s tail, which was used as a weapon against predators.

theropod
A meat-eating dinosaur that moved on two legs.

titanosaur
The biggest of the sauropod dinosaurs, such as *Argentinosaurus*.

tropical
Hot, wet conditions or places. Most rain forests are tropical.

tundra
Cold northern areas of land with frozen soil and no trees. Some plants grow here. Animals such as Arctic hares live on the tundra.

vertebrate
An animal with a backbone, such as a mammal or bird.

warm-blooded
Mammals are warm-blooded. Their body temperature stays the same whatever the temperature of their surroundings.

PRONUNCIATION GUIDE

Acanthostega : **ack-AN-tho-stega**
Acrocanthosaurus: **ACK-roe-kan-tho-SORE-us**
Albertosaurus: **al-BURT-oh-SORE-us**
Allosaurus: **AL-oh-SORE-us**
Altirhinus: **al-tih-RYE-nus**
Ammonite: **AM-oh-nite**
Ankylosaurus: **an-KIH-loh-SORE-us**
Ankylosaurus: **an-KIH-loh-SORE-us**
Archaeopteryx: **ARK-ee-OPT-er-ix**
Archelon: **AR-kee-lon**
Argentinosaurus: **AR-juhn-TEE-no-SORE-us**
Bamibiraptor: **BAM-bih-rap-tor**
Borealopelta: **bore-ee-al-oh-PEL-ta**
Caudipteryx: **kaw-DIP-tuh-riks**
Ceratosaurus: **keh-RAT-oh-SORE-us**
Chasmosaurus: **KAZ-moh-SORE-us**
Citipati: **CHIT-ee-PAH-tee**
Compsognathus: **komp-sog-NAY-thus**
Compsognathus: **komp-sog-NAY-thus**
Confuciusornis: **kon-FEW-shuh-SORE-nis**
cretoxyrhina: **kreh-tox-ee-RYE-nah**
Cryolophosaurus: **CRY-oh-loff-oh-SORE-us**
Cyamodus: **SYE-oh-moh-dus**
Deinocheirus : **dye-no-KY-rus**
Deinonychus: **dy-NON-ik-us**
Deinosuchus: **dy-no-SOO-kus**
Dilophosaurus: **dy-LOFF-oh-SORE-us**
Dimetrodon: **dy-MET-roh-don**
Diplocaulus: **dip-loh-COW-lus**
Diplodocus: **dip-LOH-doh-kus**
Edmontonia: **ED-mon-TONE-ee-ah**
Elasmosaurus: **el-LAZZ-moh-SORE-us**
Enchodus: **EN-koh-dus**
Giganotosaurus: **jig-an-OH-toe-SORE-us**
Gigantoraptor: **jig-AN-toe-RAP-tor**
Herrerasaurus: **HER-ray-rah-SORE-us**
Coelophysis: **SEE-loh-FISE-iss**
Hesperornis: **HESS-per-OR-nis**
Ichthyosaurus: **ICK-thee-oh-SORE-us**
Iguanodon: **ig-WAH-noh-don**
Ingentia: **IN-jen-TEH-ah**
Juravenator: **juh-RAH-ve-nay-tor**
Kosmoceratops: **kos-mo-SEH-ra-tops**
Kronosaurus: **Kroh-noh-SORE-us**
Maiasaura: **MY-ah-SORE-ah**
Megalosaurus: **MEG-ah-low-SORE-us**
Microceratus: **my-crow-SEH-ra-tus**
Microraptor: **MY-kroh-rap-tuhr**
Mosasaurus: **MOH-sah-SORE-us**
Nemicolopterus: **neh-me-keh-LOP-teh-rus**
Oviraptor: **OH-vee-RAP-tuhr**
Pachycephalosaurus: **pak-ee-SEF-ah-lo-SORE-us**
Parasaurolophus: **PA-ra-sore-OL-off-us**
Pergatorius: **per-goh-TORE-ee-us**
Placodus: **PLAH-koh-dus**
Plateosaurus: **PLAT-ee-oh-SORE-us**
psephoderma: **see-foh-DER-mah**
Psittacosaurus: **SIT-ak-oh-SORE-us**
Pteranodon: **teh-RAH-no-don**
Quetzalcoatlus: **KWETS-ul-koh-AT-lus**
Saltosaurus: **SAL-tah-SORE-us**
Sarcosuchas: **sahr-koh-SOOK-us**
Mawsonia: **more-SOH-nee-ah**
Moschops: **MOSK-ops**
Sauroposeidon: **SORE-oh-pos-AY-don**
Sinosauropteryx: **SIGH-no-sore-OP-tuh-rix**
Smilodon: **SMY-lo-don**
Spinosaurus: **SPINE-oh-SORE-us**
Stenoterysis: **ste-no-TER-is-is**
Styracosaurus: **sty-RACK-oh-SORE-us**
Tapejara: **tap-uh-JAR-uh**
Therizonosaurus: **THEH-rih-ZIN-oh-sore-us**
Tikaalik: **tik-AAH-lik**
Torosaurus: **TOR-oh-SORE-us**
Torosaurus: **TOR-oh-SORE-us**
Triceratops: **try-SEH-ra-tops**
Troodon: **TROH-oh-don**
Tupandactylus: **too-pan-DACK-tih-lus**
Tylosaurs: **tie-loh-SORE-us**
Tyrannosaurus rex: **ty-RAN-oh-SORE-us REX**
Varanops: **VAH-ra-nops**
Velociraptor: **veh-LOSS-ee-rap-tuhr**

Index

Acanthostega 6
Acrocanthosaurus 46
Albertosaurus 18
Allosaurus 68, 70–1
Altirhinus 28–9
ammonites 13
Ankylosaurus 32–3, 76
Archaeopteryx 60–1
Archelon 86–7
Argentinosaurus 48–9
asteroid strikes 20–1, 22

Bambiraptor 64
birds 14, 16–17, 21, 51, 60–1, 86
Borealopelta 24
brains 31

Caudipteryx 50–1
Cenozoic 5
ceratopsians 36–9, 77
Ceratosaurus 68–9
Chasmosaurus 36
Citipati 52–3, 55
Coelophysis 11
Compsognathus 31, 58–9
Confuciusornis 17
coprolite 77
crests 17, 19, 28, 66–7, 76, 78
Cretaceous 5, 10
Cretoxyrhina 86
crocodiles 15, 32, 78, 90–1
Cryolophosaurus 66
Cyamodus 85

Deinocheirus 56–7
Deinonychus 62
Deinosuchus 15
Dilophosaurus 66–7
Dimetrodon 8–9
Diplocaulus 6–7, 8
Diplodocus 44–5
dromaeosaurs 62–3
duck-billed dinosaurs 28–9

Edmontonia 33
eggs 34, 52–3, 85, 86
Elasmosaurus 88–9
Enchodus 88
eras 4–5, 10
extinction, mass 20–1

feathers 50–5, 58, 60–5

fish 74–5, 78–9, 86, 88, 91
flight 16–17, 55, 60, 78–80
fossils 24–5, 34, 38–9, 42–3, 46, 48, 52–3, 56–7, 60, 65, 70, 77, 81–2, 90, 92

gastroliths 47
Giganotosaurus 72–3
Gigantoraptor 54

hadrosaurs 28–9, 77
herds 19, 22, 27–8
Herrerasaurus 10
Hesperornis 86
hoatzin 61
horns 18, 26, 32, 36–8, 68–9

ice ages 22–3
ichthyosaurs 12, 13
Ichthyosaurus 13
Iguanodon 26–7
Ingentia 42–3

Jurassic 4, 10
Juravenator 58

Kosmoceratops 36–7

Liopleurodon 92–3

Maiasaura 29
mammals 21–3, 50, 54, 58
Mawsonia 91
Megalosaurus 14
Mesozoic 4–5
Microceratus 39
Microraptor 16
mosasaurs 13, 20
Mosasaurus 82–3
Moschops 9

neck frills 36–9
Nemicolopterus 81
nocturnal dinosaurs 58

ornithopods 27
"ostrich dinosaurs" 56
Oviraptor 54–5
oviraptorosaurs 55

Pachycephalosaurus 40–1
paleontologists 24
Palaeozoic 4

Parasaurolophus 19
placodonts 84–5
Plateosaurus 42
plates 30, 32–3
plesiosaurs 88–9
pliosaurs 92–3
Protostega 83
Psephoderma 84–5
Psittacosaurus 34–5
Pteranodon 78–9
pterosaurs 14, 16–17, 20, 78–9, 80–1
Purgatorius 21

Quetzalcoatlus 80–1

raptors 62–3

sails 74
Saltasaurus 18
Sarcosuchus 90–1
sauropods 44–8, 70, 72–3, 81
Sauroposeidon 46–7
scutes 68, 90
sharks 86
Sinosauropteryx 58
Smilodon 22
Spinosaurus 74–5
Stegoceras 40
stegosaurs 70
Stegosaurus 30–1
Stenopterygius 12–13
Styracosaurus 36

tails 30, 32, 70
Tapejara 78
thagomizers 30, 70
Therizinosaurus 57
theropods 74–5
Tiktaalik 7
titanosaurs 48–9
Torosaurus 38–9
Torvosaurus 68
Triassic 4–5, 10
Triceratops 20, 77
Tupandactylus 16–17
turtles 83, 86–7
Tylosaurus 14–15
Tyrannosaurus rex 76–7

Velociraptor 64–5

woolly mammoths 22–3